The Young Geographer Investigates

The Earth

Terry Jennings

Oxford University Press 1989

Oxford University Press, Walton Street, Oxford OX2 6DP

Oxford New York Toronto
Delhi Bombay Calcutta Madras Karachi
Petaling Jaya Singapore Hong Kong Tokyo
Nairobi Dar es Salaam Cape Town
Melbourne Auckland

and associated companies in
Berlin and Ibadan

Oxford is a trade mark of Oxford University Press

ISBN 0 19 917087 8 (limp non-net)
ISBN 0 19 917093 2 (cased, net)

Typeset in Great Britain by
Tradespools Ltd., Frome, Somerset
Printed in Hong Kong

Acknowledgements

The publisher would like to thank the following for permission to reproduce photographs:

Aerofilms p.29 (bottom left); B & C Alexander p.35 (top); Aspect Picture Library p.11 (left), p.41 (top right); British Geological Society p.9 (left); British Museum, Natural History Museum, p.14 (top left and right); John Cleare p.9 (right), p.12 (bottom right), p.13 (right), p.15 (left), p.19, p.29 (top left), p.38 (right), p.41 (top left); Robert Estall p.41 (centre); R & S Greenhill p.35 (bottom right), p.41 (bottom right); Susan Griggs/Julian Nieman p.36 (top), Griggs/P. Woldendorp p.8 (right), Griggs/Adam Woolfitt p.29 (bottom right); Robert Harding Picture Library p.15 (top and right), p.34 (bottom), p.37 (left), p.38 (left), p.39; John Hillson Agency p.32; Impact Photo Library p.27, p.29 (top right), p.35 (bottom left); Terry Jennings p.8 (left), p.11 (right), p.12 (top and bottom left), p.14 (left), pp.16, 17, 22, 23, 31 and 36 (bottom); Magnum/Jean Gaumy p.40 (centre right); Tony Morrison p.13 (left); OUP p.35 (insert bottom); Rex Features p.10; Science Photo Library, Cover, p.1, p.3, p.18; Spectrum Colour Library p.37 (right), p.42; The Telegraph Colour Library p.4, p.28, p.34 (top), p.40 (left and bottom right).

Illustrations are by John Davis, Nick Hawken, Gary Hincks, Ed McLachlan, Ben Manchipp, Bill Sanderson and Carleton Watts.

Contents

The Earth

Hundreds of years ago people thought the world was flat. Sailors on long sea journeys were afraid they might fall off the edge. Now, however, we know the Earth is shaped like a ball. We call this shape a sphere or globe. We can see the Earth is round when we watch a ship gradually disappear over the horizon. Astronauts have often seen the round Earth from out in space.

The Earth was formed many millions of years ago. It is believed the Earth came from a huge cloud of dust and gases swirling in space. Somehow the dust and gases were drawn inwards. They formed a huge ball which glowed red hot. Later the surface of this ball cooled. It hardened into the rocky ball we now call the Earth. Steam rose from the surface of the hot Earth.

The steam formed clouds. As the air around the Earth cooled, these clouds were also cooled. They produced rain. It rained and rained for thousands of years. The water filled the hollows on the surface of the Earth and formed the oceans and seas.

Ever since it was formed, the Earth has been changing. This book describes some of these changes.

Above: Planet Earth seen from the moon

Below: the formation of the Earth

Inside the Earth

The Earth is made up of layers. The outside of the Earth is covered by a layer of rocks and soil. This is the part of the Earth we live on. It is known as the Earth's crust. This crust of rocks goes underneath the oceans and seas as well.

We have seen that once, long ago, the whole Earth was a red hot ball. The outside of the Earth has cooled, but inside it is still extremely hot. The deeper you go down below the surface of the Earth, the hotter it is. Deep down in mines it is very hot. This is because underneath the crust there is a thick layer of hotter, heavier rocks. This layer is known as the mantle. In places it is so hot that the rocks have melted and flow like treacle.

There are two more layers underneath the mantle. Together these form what is known as the Earth's core. The inner core is believed to be a solid ball of hot metal. It is made up of the metals iron and nickel. Around this inner core is the outer core. This also consists of iron and nickel. But these are so hot that they are a liquid.

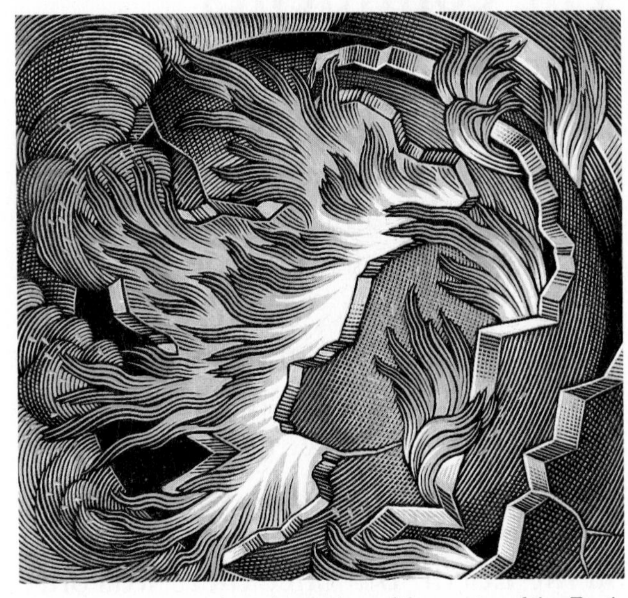

Imaginary view of the centre of the Earth

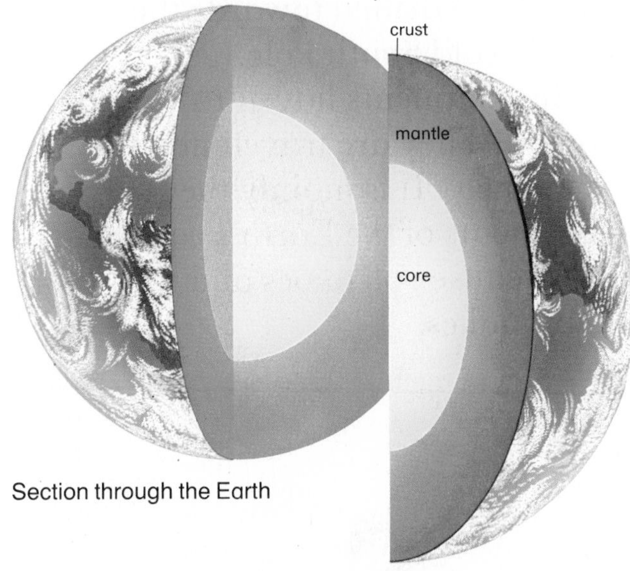

Section through the Earth

The Earth's mantle erupts to form a volcano

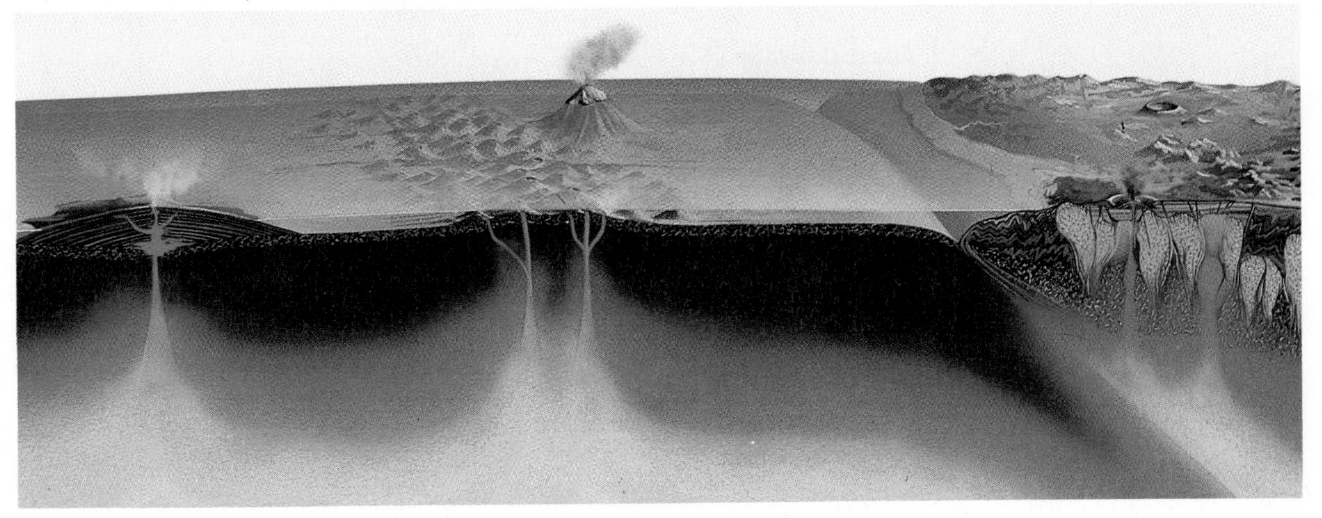

The continents

The Earth's crust is not all the same thickness. There are seven great blocks of land. These are the continents. The largest continent is Asia, the smallest Australia. On the continents, beneath mountains, the Earth's crust may be 30 kilometres thick. Under the oceans the crust is only 6 kilometres thick.

Today there are seven continents on the Earth. But once there was only one continent. Millions of years ago all seven continents were joined together. Scientists have given this super-continent the name Pangaea. Eventually Pangaea broke up. The pieces gradually drifted away from each other. They formed the seven continents we know today.

Antarctica and Australia were originally joined together as part of Pangaea. But Australia drifted slowly towards the hot tropics, while Antarctica drifted towards the bitterly cold South Pole.

The continents haven't stopped moving. They are travelling very very slowly. It is mainly the movements of the Earth's continents which cause volcanoes and earthquakes.

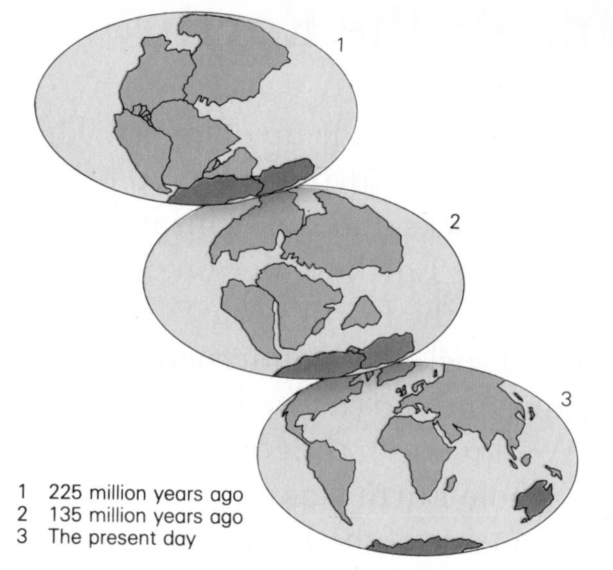

How Pangaea was formed

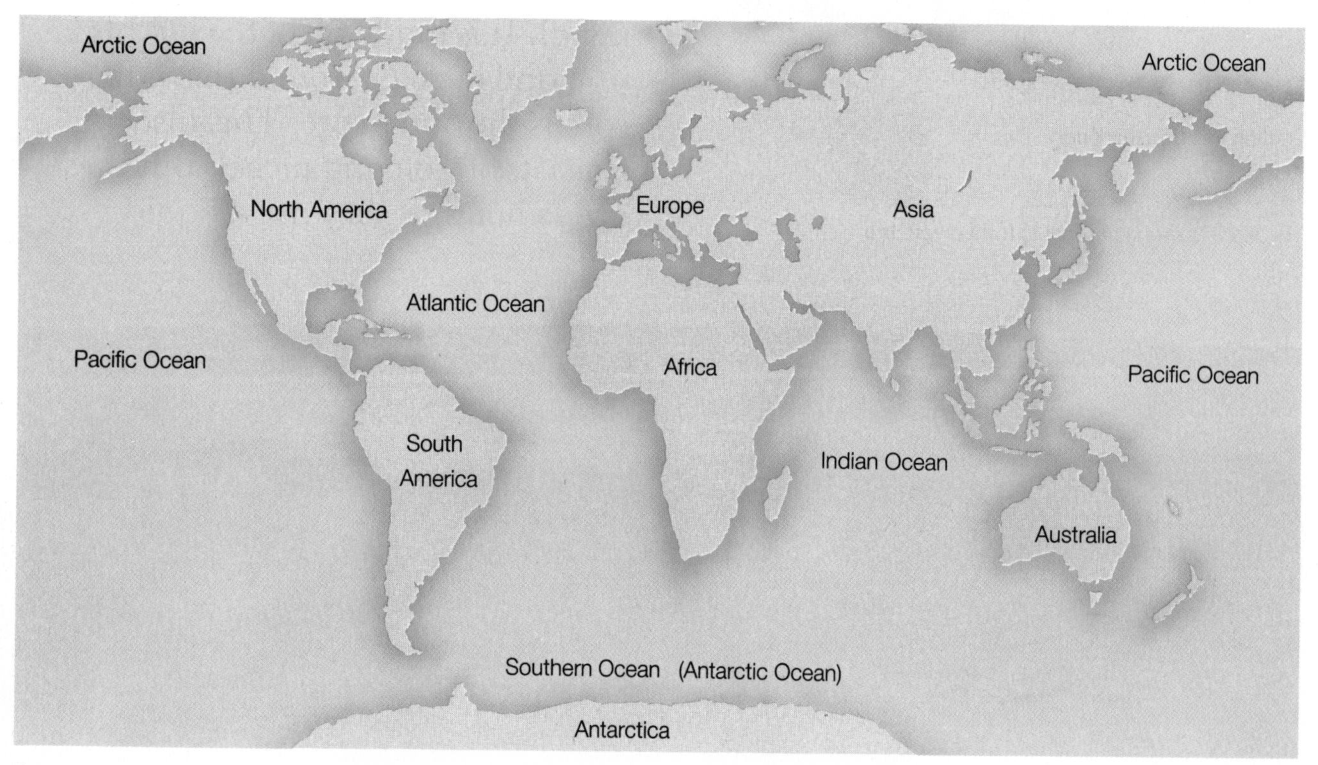

Drifting continents

The Earth's crust is not just one huge sheet of rock. It does not cover the Earth like the skin of an orange. The Earth's crust is cracked and broken into at least 15 pieces. These pieces are called plates. Some of these plates have oceans on them. Other plates carry continents.

The plates are not still. They are being slowly pushed and pulled around by movements of the hot mantle rock below them. It is the movements of the plates which cause the continents to move. North America and Europe, for instance, are slowly drifting apart. Each year they are two or three centimetres further apart. This is because a gap keeps opening up in the floor of the Atlantic Ocean between the plates bearing these two continents. Liquid rock seeps up through the gap. It hardens and keeps filling the space made as the two plates move apart. In this way a new strip of crust is always being formed in the Atlantic.

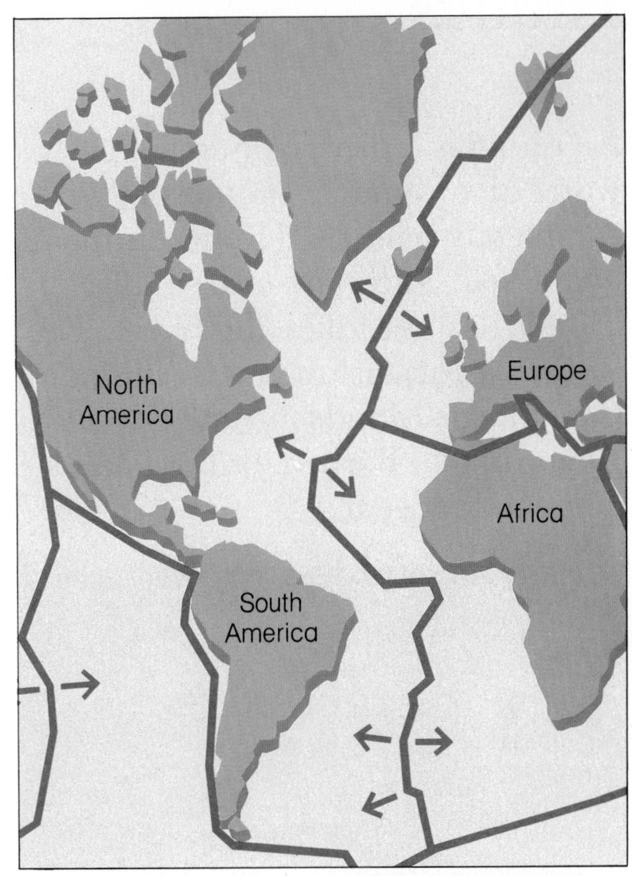

Some of the Earth's plates

In other parts of the world, the Earth's crust is being drawn down into the mantle. In the eastern part of the Pacific Ocean, for example, a narrow strip of the Earth's crust is being sucked down inside the mantle. The loss of crust here makes up for that gained in other parts of the Pacific and Atlantic Oceans.

The spreading ocean floor of the Atlantic

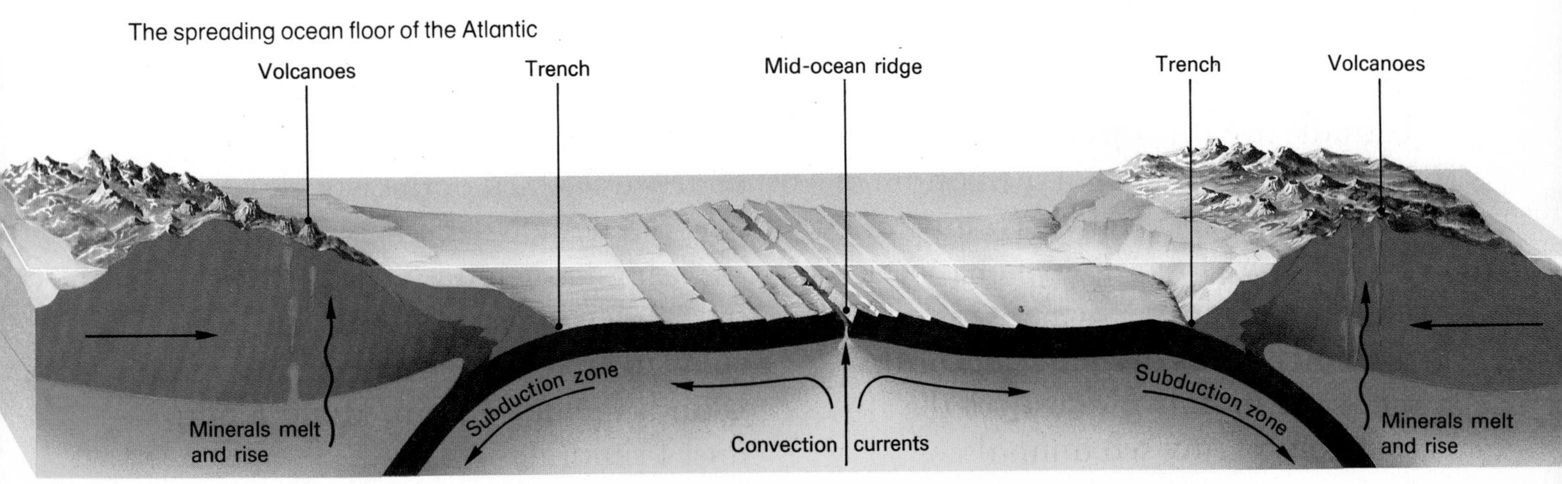

Islands

An island is a piece of land with water all around it. Islands are completely separated from continents by the sea. There are also small islands in some lakes and rivers. The world's largest island is Greenland. Other large islands include Papua New Guinea, Borneo, Baffin Island and Madagascar.

A small continental island

There are two main kinds of island. Oceanic islands lie far out to sea. Some of them, such as the Hawaiian Islands, are the tops of high mountains rising from the bottom of the sea. Others, such as Iceland and Tahiti, were built up by volcanoes under the sea. In warmer seas, many oceanic islands are made of coral. They are mainly made up of the skeletons of tiny sea animals.

Coral islands, Queensland, Australia

Continental islands always lie close to the mainland. Many of them were once joined to the mainland. They became islands because the sea level changed or because pieces of the Earth's crust moved. The British Isles were once joined to the mainland of Europe. At that time nearly one-third of the world was covered with ice and snow. The British Isles became separated when the sea level rose after this ice and snow melted. Other continental islands include Japan, Hong Kong, and Trinidad.

Faults and folds

Although most rocks are hard and strong, they will still bend and even break. You can see where the rocks have bent in the cliff in the picture.

Folds in rock

The Alps are fold mountains

These bends in the rocks are known as folds.

Sometimes, huge movements of the Earth's plates bend rock layers into giant wrinkles. The mountains formed are known as fold mountains. They include the Himalayas, Alps, Andes and Appalachians.

The formation of rift valleys and block mountains

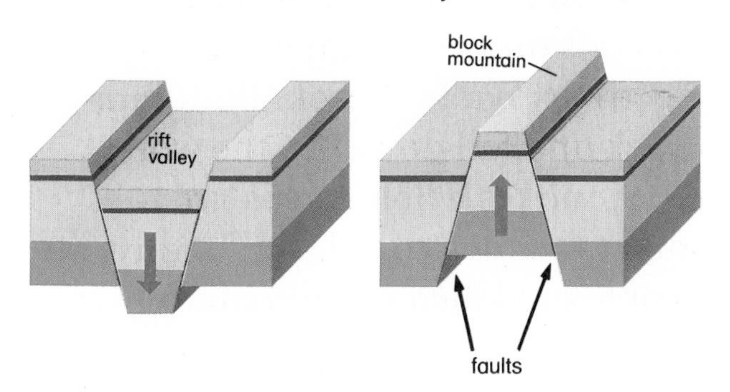

As the rocks bend they may crack or break. These cracks and breaks in the rocks are called faults. In places, parallel faults have opened up in the rocks. Huge pressures inside the Earth sometimes force up a block of land between two of these faults. The mountains formed are called block mountains. The Vosges Mountains of France and the Black Forest area of Germany were formed like this. Sometimes a block of land slips down between two faults. This produces a rift valley. One huge rift valley runs for nearly 6500 kilometres along Africa.

Earthquakes

Earthquakes are caused by movements of the Earth's plates. Sometimes the edge of one plate sticks against the edge of another. This puts such a great strain on the rocks that they bend. The rocks may also split, forming a fault. Suddenly the two plates jerk apart. They usually give way along a fault. As the plates slip apart, the land above shudders and shakes. These jerky movements of the Earth's crust are earthquakes.

Earthquake damage in Iran

In some ways an earthquake is rather like a cupboard door which closes too tightly. Pulling the door hard seems to have no effect. And then suddenly the door gives way with a thump. The person trying to open it may even fall over.

Each year about 500 000 earthquakes are recorded. Most of them are very slight and hardly noticed. But every year about 1000 earthquakes are strong enough to cause damage. Roads and railway tracks may twist and buckle. Sewers and gas and water pipes and electricity cables break. Buildings crack and fall down. In the worst earthquakes whole towns are sometimes destroyed. And many people are killed or injured.

Volcanoes

Like earthquakes, volcanoes occur where the Earth's crust is weak. Mostly this is where the plates meet. Volcanoes are made when a crack opens in the Earth's crust. Molten rock, or lava, from inside the Earth's mantle pours out. It cools to form new rock. The rocks formed by cooling the lava from a volcano are called igneous rocks. Granite, basalt and obsidian are just three igneous rocks.

When a volcano has erupted many times it may build up a tall cone-shaped mountain made of ashes and igneous rocks. Mount Kilimanjaro in Tanzania is such a volcanic mountain. So is Mount Etna in Italy.

Mount Kilimanjaro in Tanzania

Section through a volcano

However, not all volcanoes are tall mountains. In some the lava is runny and it just seeps out from holes or cracks in the ground. It then forms thick sheets of igneous rocks. Mauna Loa in Hawaii has a low cone made of sheets of rock like this.

Altogether there are about 450 active volcanoes on land. But volcanoes do not go on erupting for ever. Some volcanoes go for many years without erupting. They are then said to be dormant or sleeping. Some other volcanoes have finished erupting. They are said to be extinct. Edinburgh Castle in Scotland and the chapel at Le Puy in France are built on extinct volcanoes.

An ancient extinct volcano on Dartmoor, England

Mountains are being worn down

Scree slopes on a mountain

Even as mountains are being formed by volcanoes and movements of the Earth's plates, they are being worn away. This gradual wearing away of rocks happens in several ways.

All rocks have tiny cracks in them. When rain falls or dew forms, water gets in these cracks. If it freezes, the water turns to ice. As water turns to ice it expands or gets bigger. The ice presses hard against the sides of each crack. Eventually the rocks weaken and pieces break off them.

In hot, dry places such as deserts, hot days and cold nights weaken the rocks. Slowly the rocks break up into smaller pieces. Sand blown by the wind may gradually wear rocks away. Pieces broken off hills and mountains slide down the sides. They may form a loose bank of stones called scree. Rainwater may also rot some rocks. Rainwater is a very weak acid. It gradually dissolves limestone.

Ice, heat, cold, wind and rainwater, which break down rocks, are all examples of what is called weathering. In addition, plant roots grow in cracks in rocks. As the roots grow they force open the cracks. Again, pieces may be broken off the rocks.

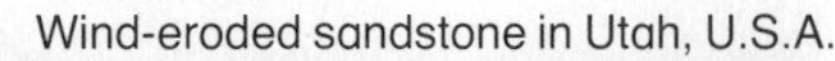

Weathered limestone rocks on Minorca

Wind-eroded sandstone in Utah, U.S.A.

River and glaciers

Rivers make many changes to the surface of the Earth. Most rivers start as springs high up on hills or mountains. The water comes from rain which has soaked into the ground. The stream coming from the spring flows quickly downhill. As it flows, the water sweeps away small pieces of weathered rock. These small pieces of rock rub on the bottom and sides of the stream, deepening and widening it. They gradually carve out a steep-sided V-shaped valley.

As the stream flows along it is joined by other streams. Together they form a small river which grows as it is joined by more rivers and streams. In time the river makes its valley deeper and wider. As the valley widens, its sides get less steep and the slope of its bed is gentler. The river begins to wind or meander along. It still wears away the outer bends but it drops some of the small pieces of rock on the inside of the bends.

Near its mouth the river has often completely worn away the hills and mountains. It flows slowly towards the sea over a broad, low plain.

Big rivers drop huge loads of mud and sand on the sea-bed around their mouths. This mud and sand is formed mainly by water grinding up small pieces of rock the river has carried along. In the colder parts of the world there are rivers of ice called glaciers. These too wear away the rocks as they move along. The valleys they carve are U-shaped.

Stages in the life of a river

A glacier in the Rocky Mountains, U.S.A.

The mouth of a river

New rocks from old

Mud, sand and stones, are carried by rivers down to the sea. They settle, or sediment out, into thick layers on the bottom of the sea. The mud, sand and stones broken off rocks by the sea may also pile up on the sea-bed. After thousands of years they are pressed into new rock.

Limestone, chalk, sandstone and most kinds of clay were made at the bottom of the sea. These rocks are called sedimentary rocks because they were made from sediments. They were formed in layers called strata. You can often see these strata in cliffs and the sides of quarries.

Rock layers or strata in a sea cliff

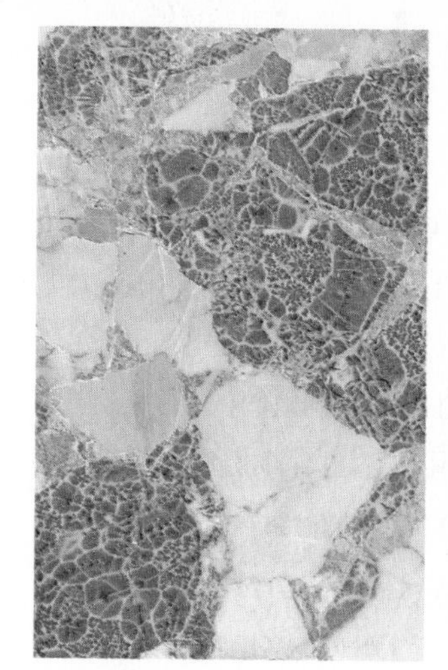

Examples of marble (left) and slate

Sometimes sedimentary rocks are later changed by heat, by chemicals and by other rocks pressing down on them. The igneous rocks produced from volcanic lava can also be changed in these ways. All kinds of rocks can be changed. Slate which is sometimes used for roofs was once soft shale. The beautiful rock called marble, which is used for fine buildings, statues and gravestones was once limestone. Rocks such as marble and slate which have been changed are called metamorphic rocks. The word 'metamorphic' means 'changed'.

Materials and fuels from the Earth's crust

Open-cast coal-mine in Australia

We use rocks for many things. Some buildings are made directly from rocks. The cement and concrete used in buildings are also made from rocks. Small pieces of granite and other hard rocks are used to surface roads. However, not all rocks are hard. One of the best-known rocks is clay. Clay is used to make bricks, pottery and sculpture.

A potter at work in Nepal

A rock is made up of chemical substances called minerals. Some of these minerals contain metals. Most metals are found in the ground in mixtures known as ores. The metals can be obtained from these ores. Many minerals occur as beautiful shapes called crystals. Some of these mineral crystals are very rare. They are known as precious stones. The rarest and most beautiful precious stones are diamonds, emeralds, rubies and sapphires.

Our important fuels also come from rocks. Coal is the remains of trees and other plants which grew in forests millions of years ago. Oil is another fuel found in the Earth's rocks. It was made from tiny sea creatures that lived millions of years ago. There is often natural gas with the oil, and this too can be used as a fuel. The uranium which is used as a fuel in nuclear power stations is obtained from certain rocks. Building materials, pottery, minerals, metals, precious stones and fuels are just a few of the things we obtain from the Earth's crust.

Building materials such as steel, concrete and glass come from the Earth's crust

Soil

Potting begonias in garden soil

Almost everywhere on land there is soil. Sometimes there is a lot of soil. In rocky places and deserts there is often very little soil. It is too dry in these places for plants to grow and help to form soil. Even when soil has been formed it may be too poor for most plants to grow in it. Other soils are good and grow many plants. A good soil which grows many plants is said to be fertile.

Soil is very valuable to us. All the plants we eat grow in soil. So do the cotton and flax plants which provide us with some of our clothing. The trees which give us our timber and paper grow in the soil. Our meat and milk come from animals which feed on plants that grow in soil. Wool and leather also come from these animals.

Without soil we would have nothing to eat or read, nothing to write on, and little to wear. Nor would we have any of the things which are made from timber.

Fertile oak woodland in Norfolk, England

How soil is formed

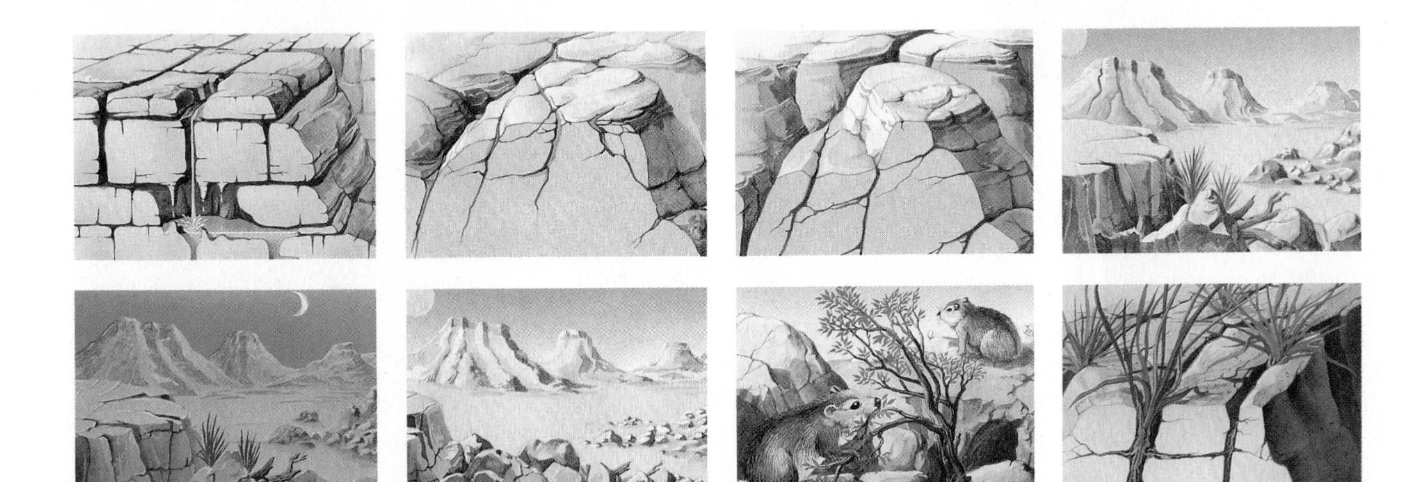

Soil is slowly being formed all the time. All our soil was made from rocks. Frost and plant roots crack rocks. The sun's heat by day and the cold at night breaks rocks. Streams and rivers wash pieces of rock away. The sea batters rocks and breaks them up. Small pieces of sand blown by the wind scratch rocks and break pieces off them. As we have seen, these pieces of rock may be carried by the wind, rivers, streams and the tides to other places. In some countries, rivers of ice called glaciers break off pieces of rock and push them along.

If the pieces of rock stay on land, small plants may grow in the crevices between them. Small plants may also grow on the surface of the rock. Mosses and lichens can often be seen growing on pieces of rock. When these simple plants die, bacteria and moulds grow on them. The dead plants decay and break up into little pieces. The decayed pieces of plant are food for other plants and also for animals. Eventually the tiny pieces of rock have lots of black pieces of decayed plants and animals mixed in with them. The black substance is called humus. When there is a lot of humus mixed in with tiny pieces or grains of rock, a soil has been formed.

Plants growing in a crack in rock

Lichens growing on a boulder

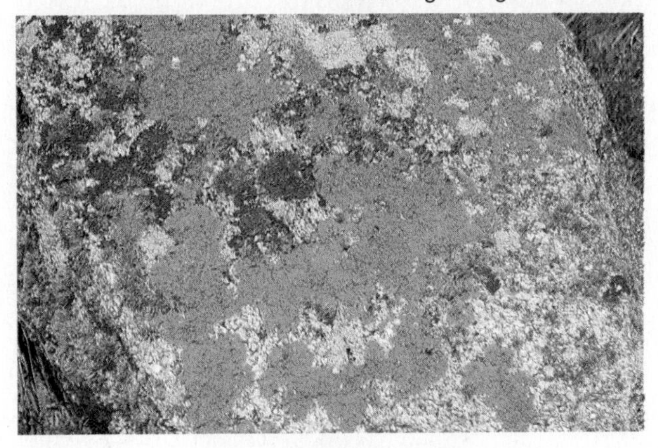

Day and night

Day and night on the Earth, as seen from space

All the time our Earth is spinning round. It spins as if it was a giant top turning on an invisible axis. The upper end of this invisible axis is the North Pole. The bottom of the invisible axis is the South Pole.

The sun only shines on half of the world at a time. The half of the Earth facing the sun has day. As the Earth turns, the places on this half of the Earth gradually turn away from the sun. They go into the area of shadow. We call this period of dark shadow, night. At the same time, places where it was night slowly turn towards the sunlight. A new day begins to dawn in these places.

The Earth makes one complete turn every 24 hours. But not all places have 12 hours day and 12 hours night. This is because the Earth is tilted on its axis. On the next page we shall see some of the other effects this tilt of the Earth has.

Why we have days and nights

The seasons

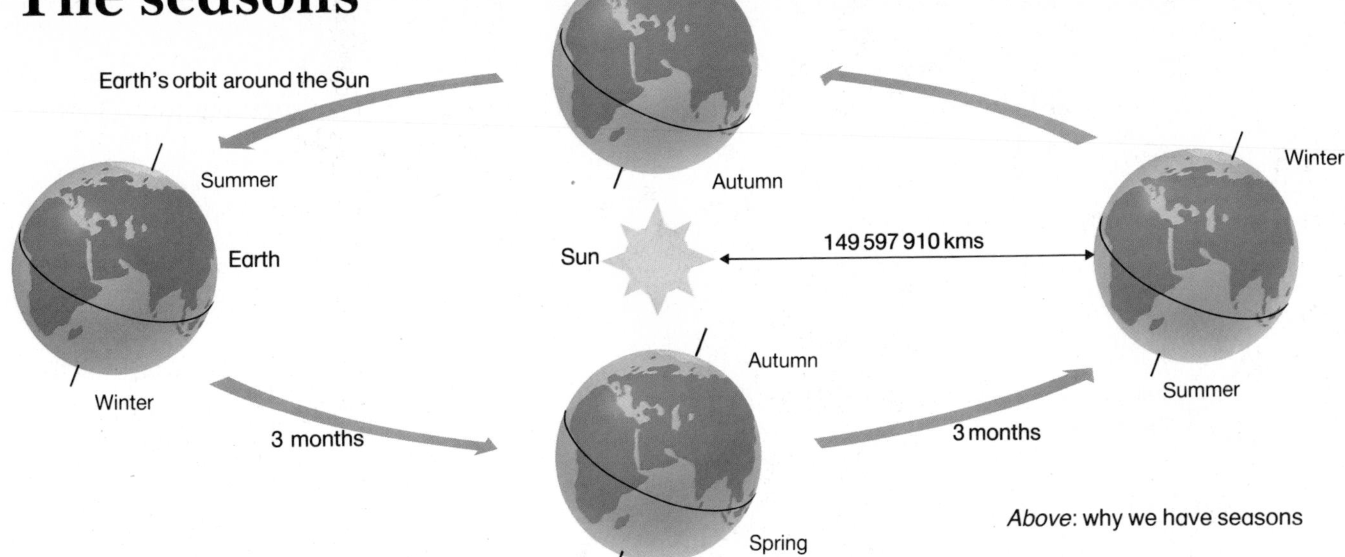

Above: why we have seasons

Not only does the Earth spin on its own axis, it also moves around the sun. The Earth follows a curved path or orbit around the sun. The Earth takes 365¼ days to travel around the sun. That is why every fourth year an extra day is added to the month of February. The so-called leap year keeps the calendar correct.

As we have seen, the Earth is slightly tilted on its axis. It is this tilt of the Earth which gives us our seasons. For some months the North Pole is tilted towards the sun. The northern half or hemisphere of the Earth has summer. The days are longer and the temperatures higher in summer. The southern hemisphere tilted away from the sun is colder. The sun's rays have further to travel and they lose more of their heat as they pass through the air. The southern hemisphere has winter.

Six months later the Earth is on the opposite side of the sun. The North Pole leans away from the sun and the northern hemisphere has its winter. The southern hemisphere, being tilted towards the sun, has its summer. Spring and autumn are in the in-between seasons.

For at least a few days in the middle of summer the sun never sets in the polar regions. Because of this, the areas around the North and South Poles are often called the 'lands of the midnight sun'. In the winter both polar regions are extremely cold. In mid-winter the sun does not rise above the horizon and it is dark all day and all night.

Below: midnight in summer in Antarctica

Do you remember?

1 What do we call the shape of the Earth?

2 From what was the Earth formed?

3 How were the oceans formed?

4 What is the Earth's crust?

5 What is the Earth's mantle like?

6 What two metals make up the Earth's core?

7 How many continents are there?

8 What do we call the 15 or so pieces which make up the Earth's crust?

9 What is an island?

10 Name the two main kinds of island.

11 What is the proper name for bends in rocks?

12 What are the large cracks and breaks in rocks called?

13 What is humus?

14 How are earthquakes caused?

15 Where do volcanoes occur?

16 What is lava?

17 What are igneous rocks?

18 Describe three ways in which rocks are broken down.

19 What shapes are the valleys made by a) fast-flowing rivers, b) glaciers?

20 What are meanders?

21 How are mud, sand and stones sometimes turned into new rocks?

22 What are metamorphic rocks?

23 What is an ore?

24 Name three fuels which we get from the Earth's crust.

25 What do we mean when we say a soil is fertile?

26 Name five of the things we obtain directly or indirectly from the soil.

27 How are block mountains formed?

28 How long does it take for the Earth to make one complete turn on it axis?

29 What season does the northern hemisphere have when the North Pole is tilted towards the sun?

30 What season does the southern hemisphere have when the South Pole is tilted towards the sun?

Things to do

1 A model of the centre of the Earth Use different coloured pieces of plasticine to make a model of a section through the Earth, like that in the picture on page 5. On your model show the inner and outer cores, the mantle and the Earth's crust.

2 Write a story Write a story called 'Journey to the centre of the Earth'.

3 A model of Pangaea Obtain or draw a small map of the world. Stick your map onto a thin sheet of polystyrene plastic, such as a polystyrene ceiling tile. Use a sharp knife (Careful!) to cut out the seven continents. Arrange them together so that they form a model of Pangaea (see page 6). Which continent had to move the greatest distance to reach its present position? Which continent had to move from a warm or hot climate to a very cold one?

4 Folds and faults Use layers of different coloured clay, plasticine or papier mâché to make models of folds and faults in the Earth's crust.

5 Make a model earthquake Find two planks of wood both the same thickness. Lay them end to end. Use Lego or some other building blocks to make model houses. Place these across your 'fault'. Move the two planks quickly either horizontally, vertically or both. Watch your buildings topple.

Can you make model buildings which will withstand a small model earthquake?

6 Volcanoes Whereabouts in the world are volcanoes found? Mark the places on a map of the world. Do the marks make a pattern?

Whereabouts in the world do earthquakes happen? Mark these on the map as well. Do the marks showing where earthquakes happen make a pattern? Do earthquakes happen in the same parts of the world as volcanoes?

7 Make a model volcano Make the cone of the volcano with plasticine or plaster of Paris. Do not forget to make a hole in the top of the volcano for the crater.

Take a tuft of cotton wool. Paint it red, orange and black. Stick it in the crater of the volcano. The cotton wool can be the smoke and flames coming from the volcano.

8 Collect stamps Make a collection of postage stamps which show pictures of mountains and volcanoes. Display your stamps in an album or on a wallchart. Write a sentence or two about each of the stamps, the mountain or volcanoes they show, and the countries the stamps came from.

9 What happens when water freezes? This simple experiment will help you to understand what happens when water freezes in rock cracks.

Completely fill a small plastic bottle with water and screw the top on tightly. Put the bottle in the freezing compartment of the refrigerator or in the deep-freeze overnight. What has happened to the bottle the next day? Why is this?

Put a small piece of sandstone, chalk or limestone in a bowl of water overnight. Then place the piece of rock on a tin lid and stand it in the freezing compartment of the refrigerator or in a deep-freeze overnight. See what has happened the next day. Why is this?

Try this experiment with other kinds of rocks and building materials, including small pieces of concrete and brick.

Can you think why, after a really hard winter, the roads and pavements sometimes have large pot-holes in them?

A weathered stone cross on Dartmoor, England

A weathered limestone wall

10 Weathering Look to see how the weather has affected gravestones. Compare weathering over different periods of time (remember that gravestones have the dates on!). Which kind of rock wears best over a long period of time?

Look at walls and statues to see how they are affected by the weather. Which building materials withstand the weather best?

Do gravestones and walls weather more quickly in the town or in the country? Why is this?

11 Make a model mountain and river scene If you can obtain permission, make a mudpie mountain in the garden. Pat the mud down smooth. Push some stones into the mud and then let it dry.

Water your mountain with water from a watering can. Where does the water wash away, or erode, your mountain? Watch how the water forms miniature streams, rivers and waterfalls. Are any valleys formed? What shape are they? Where does the mud which is eroded away from the mountain finish up?

What happens if you make a mudpie mountain and cover it with grass plants? Does the mud wash away now? What effect do the grass plants have?

12 A model river Use clay, plasticine or papier mâché to make a model of the course of a very *short* river. Remember that rivers begin on higher land, flow in valleys and end at the sea. Paint your model and display it in your classroom. Write a story about the life of your river.

13 Make a collection of rocks

Anyone can do this, not just those who live near mountains.

Look for pebbles and pieces of rock in the garden, at the seaside, by the roadside and in the country. River banks, the beds of streams, road cuttings, old quarries and building sites can also produce rock samples. These last places are *dangerous*, though. Do not go to them without permission, and *always* go with a grown up.

Wash your rocks and pebbles carefully and dry them. Label each one with the name of the place where you found it and the date. Make a display of your rocks and pebbles for your friends to look at.

When you are searching for rock specimens, always keep a lookout for fossils. Make a collection of these as well. Carefully clean and wash each one. Label it saying where and when you found it.

Use books to try to find the names of your rocks and fossils. Try to find out how each kind of rock in your collection was formed.

14 Visit your local museum

Look at their collection of rocks, minerals and fossils. How many of the rocks, minerals and fossils were found locally?

You may be able to find out the names of some of your own rocks and fossils if you compare them with those in the museum's collection.

15 Grow crystals

As we have seen, when the lava from a volcano cools down, often beautiful shapes called crystals are formed.

You can make some crystals of your own if you add salt to some water in a clean jam-jar. Stir the salt until it dissolves. Keep adding salt to the water until no more will dissolve. You have now made a strong or concentrated salt solution.

Pour a little of the salt solution into a clean saucer and leave it on a windowsill. When all the water has evaporated, look with a hand lens or magnifying glass at the crystals left in the saucer. What do you notice about them?

If you want to make big crystals, evaporate your salt solution very slowly.

Make some more crystals in the same way. Try sugar, washing soda, alum, lemonade powder, or Epsom salts. Alum and Epsom salts can be bought quite cheaply from a chemist's shop.

16 Collect pictures

Collect pictures of all the wonderful scenery in the world in which rocks can be seen. Include hills, mountains, river valleys, waterfalls, cliffs, beaches, caves, deserts, volcanoes and glaciers. Make a book or wallchart with your pictures. Label each picture saying what it shows and where the picture was taken.

17 Metals

Find out the names of as many metals as you can. Write their names across a sheet of paper and then draw columns underneath them.

Find out as many things as possible which are made of these different metals. Write their names in the columns under the names of the metals.

Which countries do the different metals come from? How are the metals obtained from the ground? What effect, if any, does obtaining the metals have on the surrounding countryside?

What substances are sometimes used instead of the metals nowadays? Why?

gold	silver	iron and steel	copper	aluminium	zinc
jewellery					
filling teeth					
gold letters on books					
coins					

18 What soil is made of Put a handful of garden soil in a jar with straight sides. A clear-glass coffee jar is a good one to use. Then pour in water until the jar is about three-quarters full.

Put the lid on the jar and shake it really well. As soon as you have stopped shaking it, put the jar on a table or windowsill and watch carefully to see how the soil settles.

How long does it take for the water to clear? Can you see the layers of the different sized pieces of soil? How many layers can you see?

Which layer has the largest pieces? Which layer has the smallest pieces? Measure the thickness of each of the layers.

Are there any little pieces of humus and decaying plants and animals floating on the top of the water? These usually look black.

Do the same thing with soils from other places. If you always use the same jar you can make a table to compare the thicknesses and colours of the different layers in the different soils. You could also make drawings of the jars containing each of the different soils.

19 How does the kind of soil affect the way in which plants grow? Dig a hole in the garden about 30 to 40 centimetres deep. Fill a plant pot or yoghurt pot with soil from the bottom of the hole, and another one from the surface of the ground.

Look at the two soil samples with a hand lens or magnifying glass. What colour are they? What do they feel like? What do they smell like? Are there any other differences between the two samples of soil?

Sprinkle cress seeds on the soil in both pots. Gently press the seeds into the soil with your fingers or with a small piece of flat wood.

Keep both pots watered regularly, using the same amount of water for each pot.

In which pot do the seeds grow best? Can you think why this might be?

If gardeners and farmers plough or dig very deeply, will they grow better crops?

Fill some more pots with the soils from different places. Sprinkle a few cress seeds on to the surface of the soil. Gently press the seeds into the soil.

Keep all the pots watered regularly, using the same amount of water for each pot.

In which kind of soil does the cress grow best? Draw a picture to show what the little cress plants in each pot look like.

20 Make a soil profile Find a deep tin or jar lid. Roll a sheet of acetate or thin, clear plastic into a tube which fits neatly into the lid. Seal the acetate or plastic into a roll with Sellotape.

Ask permission to dig a hole in a part of the garden which hasn't been dug for a long time. Dig a hole which is as deep as your tube is long. Dig the hole carefully and clean the sides of all loose soil.

Put a little of the soil from the bottom of your hole in the bottom of the tube. Then take some of the soil from a few centimetres above the bottom of your hole and put that in the tube. Do this until your tube shows exactly how the soil in the sides of your hole changes.

Look at the soil profile in the tube carefully. Where is the darkest soil to be found? Where is the lightest soil? Where are the most stones? How far down do plant roots grow?

Compare your soil profile with those from your friends' gardens. Are they all the same?

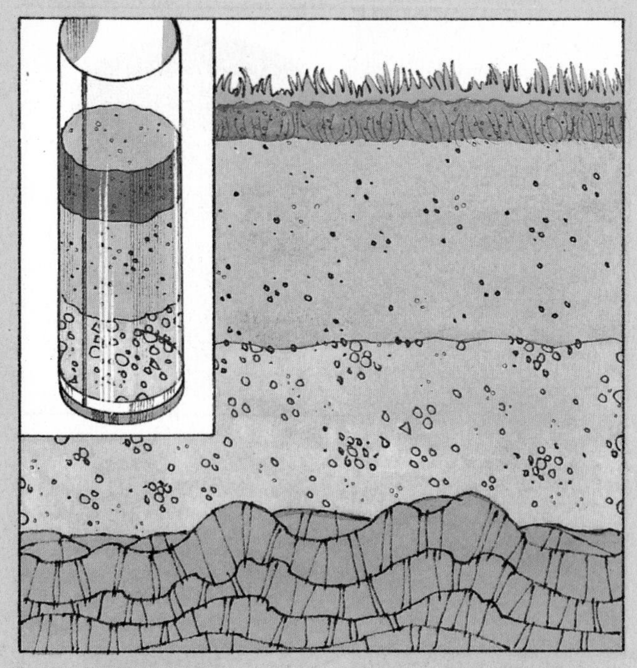

21 Night and day Obtain a globe and a desk lamp or a powerful torch. If possible work in a room with the curtains drawn.

Shine the light from the desk lamp or torch on to the globe from the side. Turn the globe slowly in an anticlockwise direction. See how each country gradually moves into the light. Notice where it is dawn, where it is day, where it is dusk, and where it is night.

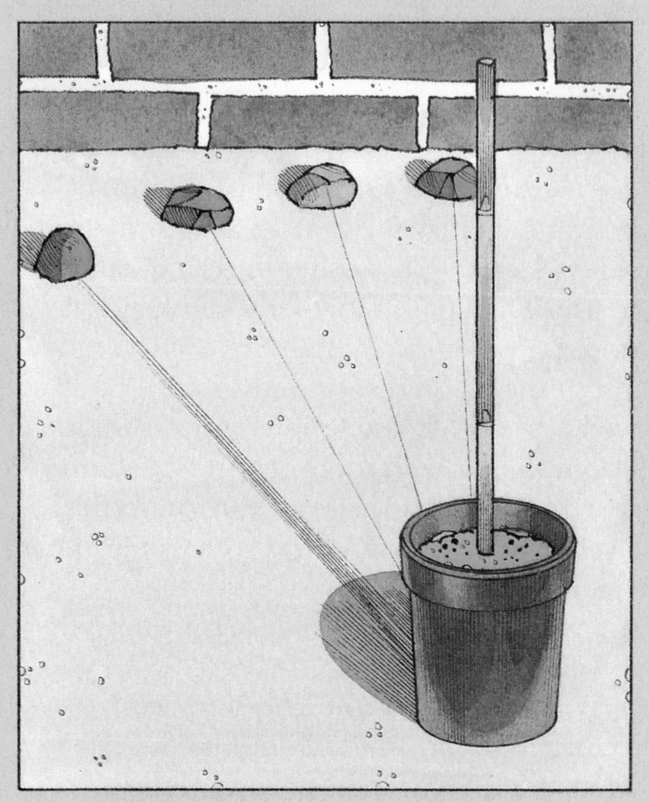

22 Make a shadow clock Push a stick upright into the lawn. Every hour on a sunny day, measure the length of the shadow made by your stick. Each hour mark the end of the shadow with a small stone. When do you see the shortest shadow? When do you see the longest shadow? Why does the shadow move?

If you are not able to put a stick into the lawn, you could always push a stick into a pot of soil standing on the playground, as shown in the picture.

If you had been locked out of doors on a sunny day and did not have a watch, how would you tell when it was midday?

Try making the shadow clock on a sunny day at another season of the year. Do you get the same measurements?

Things to find out

1 The Earth is a planet. It is one of nine planets which orbit around the sun. Find out the names of the other eight planets.

2 Why do you think it is that we cannot feel the Earth spinning round? Find out why it is we do not fall off the Earth and drift out into space.

3 How is it that scientists know so much about the centre of the Earth, even though no one has ever drilled down as far as the mantle?

4 How many ways can you think of in which we depend on the heat or light from the sun?

5 Draw or trace a map to show the mountainous areas of your country.

6 Can you find any evidence of past or present volcanic activity near to where you live?

7 Why would you not expect to find fossils in igneous rocks?

8 Why are granite, limestone and sandstone used for quite different purposes? What makes each one particularly suitable for these different purposes?

9 Sometimes after a big volcanic eruption there may be spectacular sunsets. The summers may be cooler and wetter and the winters colder. Find out what causes these changes in the weather.

10 Make a list of all the things people use rocks for. How many things can you find?

11 Coal is sometimes called 'stored-up sunlight' or 'bottled sunshine'. Find out why this is.

12 What is meant by latitude and longitude? How are they used on maps?

13 Find out, or record, how many hours of daylight there are in each of the months of the year. What else can affect the amount of daylight besides the position of the sun in the sky?

14 Find out who invented calendars. Who introduced the calendars we use today? Why are calendars useful?

15 Can you find out how the months of the year obtained their names?

16 What differences might you notice between a sunny day in summer and a sunny day in winter?

A sunny day in summer (left) and winter (right)

Oceans and seas

If you look at a globe or a map of the world, you will see that there is much more water than land. Nearly three-quarters of the world is covered by oceans and seas.

There are four great oceans. These are the Atlantic, Pacific, Indian and Arctic Oceans. Some people consider that there is also an Antarctic or Southern Ocean. The largest ocean in the world is the Pacific Ocean. It covers about a third of the surface of the Earth. Its average depth is over 4000 metres. In spite of their vast size, all the oceans are connected to each other. This means that their waters are mixed together.

As well as the oceans, the surface of the Earth is covered by a number of seas. Some of these, such as the Arabian Sea and the Sargasso Sea, are parts of oceans. Other seas are surrounded by the continents. They are separate from the oceans. The Mediterranean Sea, Red Sea and Black Sea are all surrounded by continents. Some shallow seas, such as the North Sea and Baltic Sea are the flooded edges of continents. The largest of the world's seas is the South China Sea.

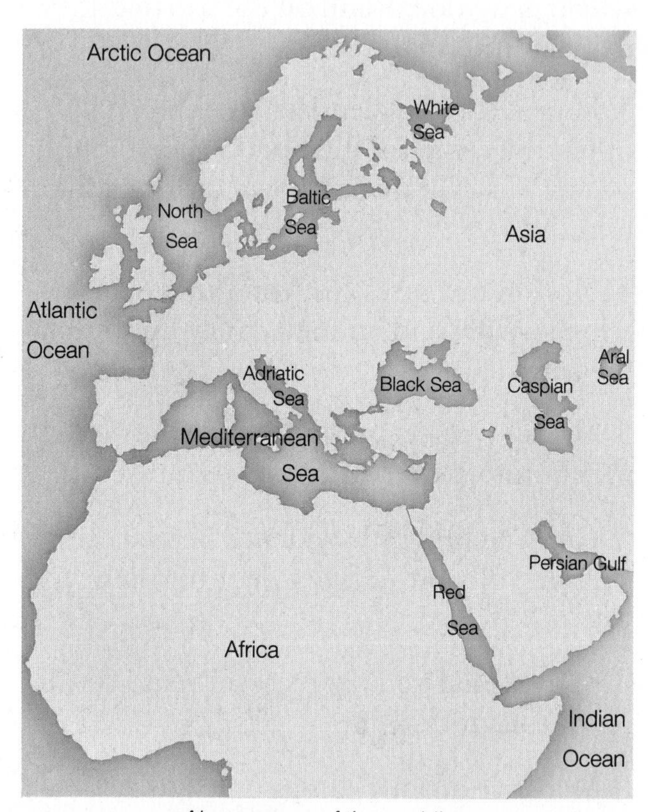

Above: some of the world's oceans and seas

Below: view from a satellite of Europe and part of North Africa

Sea water

The water in seas and oceans is always salty. Some of this salt has come from volcanoes under the sea. But most of it has come from rocks on land. When rain falls, it dissolves some of the salt in the rocks. Rivers and streams dissolve even more of the salt. They carry it down to the sea.

The sea under a sunny sky (left) and dull sky (right)

Warm seas are much saltier than cold ones. This is because the hot sun evaporates some of the sea water. The salt is left behind. And so, over millions of years, these seas have become saltier and saltier. The Mediterranean and Red Seas are very salty. But the saltiest sea in the world is the Dead Sea. The Dead Sea is so salty that no plants or animals can live in it. If people swim in the Dead Sea, the salt in the water stops them from sinking. In some parts of the world salt is obtained by evaporating sea water.

Floating in the Dead Sea

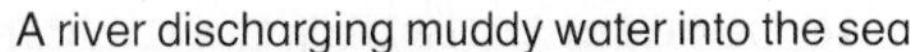

A river discharging muddy water into the sea

We often say the oceans and seas are blue. But sea water is usually colourless. Sea water looks blue because it reflects the sky. On dull, cloudy days the sea looks grey. Some seas are coloured by mud. The Yellow Sea gets its colour from the clay washed into it by rivers.

The restless oceans and seas

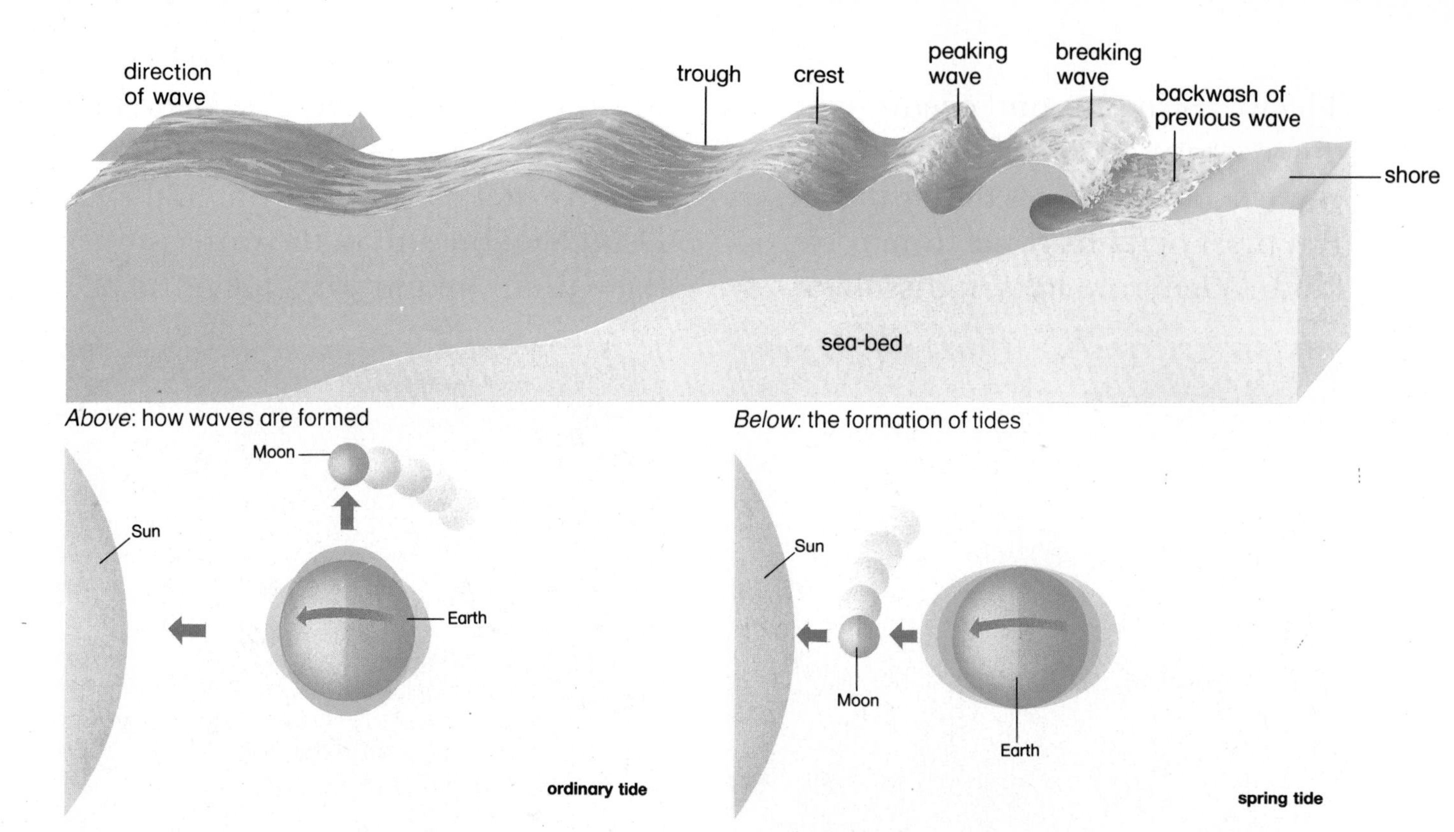

Above: how waves are formed

Below: the formation of tides

The seas and oceans are never still. Tides, waves and currents keep the water moving. Waves are made by the wind. They do not move the water from place to place like currents or tides. Waves are just big ripples of water made by the wind.

Tides are caused by several forces. The most important of these is the pull or gravity of the moon. This makes the oceans' waters pile up in a bulge on the side of the Earth which faces the moon. Another bulge is formed on the opposite side of the Earth. The highest and lowest tides are called spring tides. They occur twice a month when the sun and moon are both on the same side of the Earth. Then both the sun and moon pull on the oceans' waters and make it pile up.

As well as causing waves, winds also produce currents in the oceans. Currents are great bands of water moving through the oceans. Most currents are made by winds which blow in the same direction all the time. Some currents travel across the surface of the water. Others travel deep down. Some currents are formed because cold water is heavier than warm water. Cold water near the polar regions sinks to the bottom of the oceans making a current deep down. Warm water moves in to take its place, forming another current. One of the most important currents is the Gulf Stream. It carries warm water from near the Equator northwards along the shores of France, the British Isles, Norway and Iceland.

Under the oceans and seas

Rocks and boulders, Concarneau, France

A sandy beach in Brittany, France

A pebble beach in County Kerry, Ireland

A section through the seashore

Where the land meets the oceans and seas is the seashore. Some seashores are sandy, others are rocky, muddy or covered with shingle, pebbles or boulders. Some of these pieces of rock have come from the cliffs nearby. Others have been carried there from far inland, mainly by rivers.

Around the continents the oceans and seas are fairly shallow. At its deepest the water around the continents is only about 200 metres deep. This shallow water covers a gently sloping platform of land. It is called the continental shelf. Most of the fish we eat are caught in the rich waters over the continental shelf.

Where the continental shelf ends there is a long deep slope. This is known as the continental slope. The continental slope plunges downwards like a cliff. In some places, water and mud flowing from rivers on land have carved deep valleys in the continental slope.

The continental slope levels out into an extremely deep underwater plain. This is known as the abyss. The abyss is covered with a thick layer of slimy mud. This is mostly made up of the shells of tiny sea animals. They sank to the sea bed when they died. The abyss is dark and cold. Sunlight cannot reach these great depths. The water hardly moves.

The bottom of the ocean

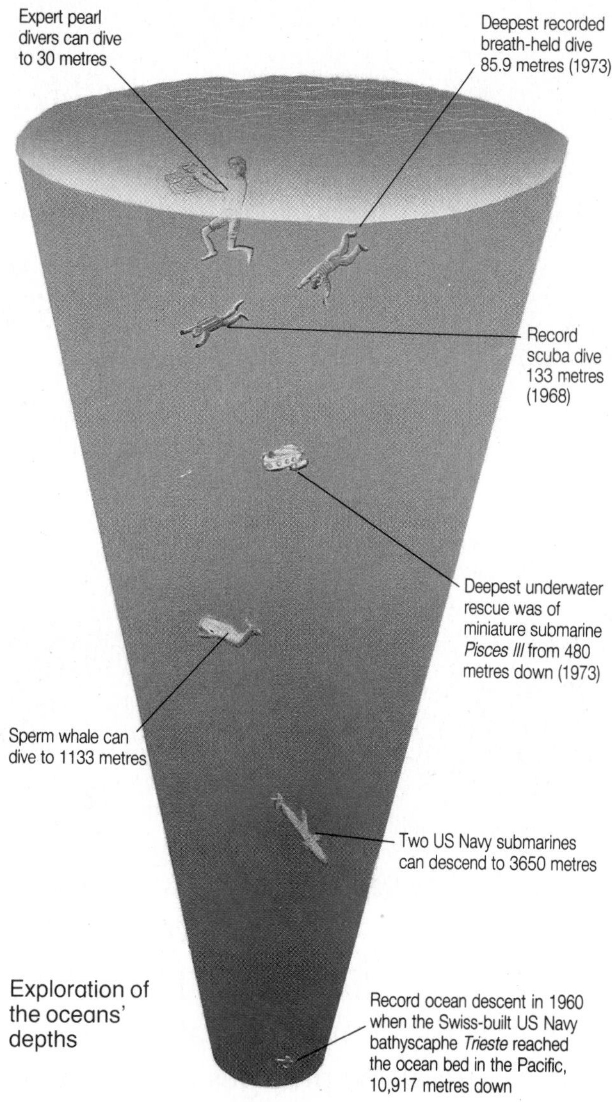

Exploration of the oceans' depths

In many ways the bottom of the ocean is like the surface of the land. There are flat, sandy plains. In places, water currents push the sand up into dunes, just as the winds on land make sand dunes in the deserts. There are mountain ranges and deep valleys on the ocean floor. Some of the mountains are so high that they rise above the surface of the water as islands. The island of Mauna Kea in Hawaii is really an undersea mountain. From its base to its peak, it is 10 023 metres high. This is higher than Mount Everest.

The mountains under the oceans and seas were formed by volcanoes on the sea-bed. Molten rock or lava comes from cracks in the sea-bed. The lava turns solid and builds up into mountains. There is a huge mountain range down the middle of the Atlantic Ocean. This was built of lava from undersea volcanoes.

There are also deep trenches under the oceans. These are massive cracks which have opened up in the sea-bed. The Pacific Ocean, over the Mariana's Trench, is more than 11 000 metres deep. A large stone dropped into the Pacific over the Mariana's Trench, would take more than 60 minutes to reach the bottom.

A submersible craft for exploring the ocean depths

The atmosphere

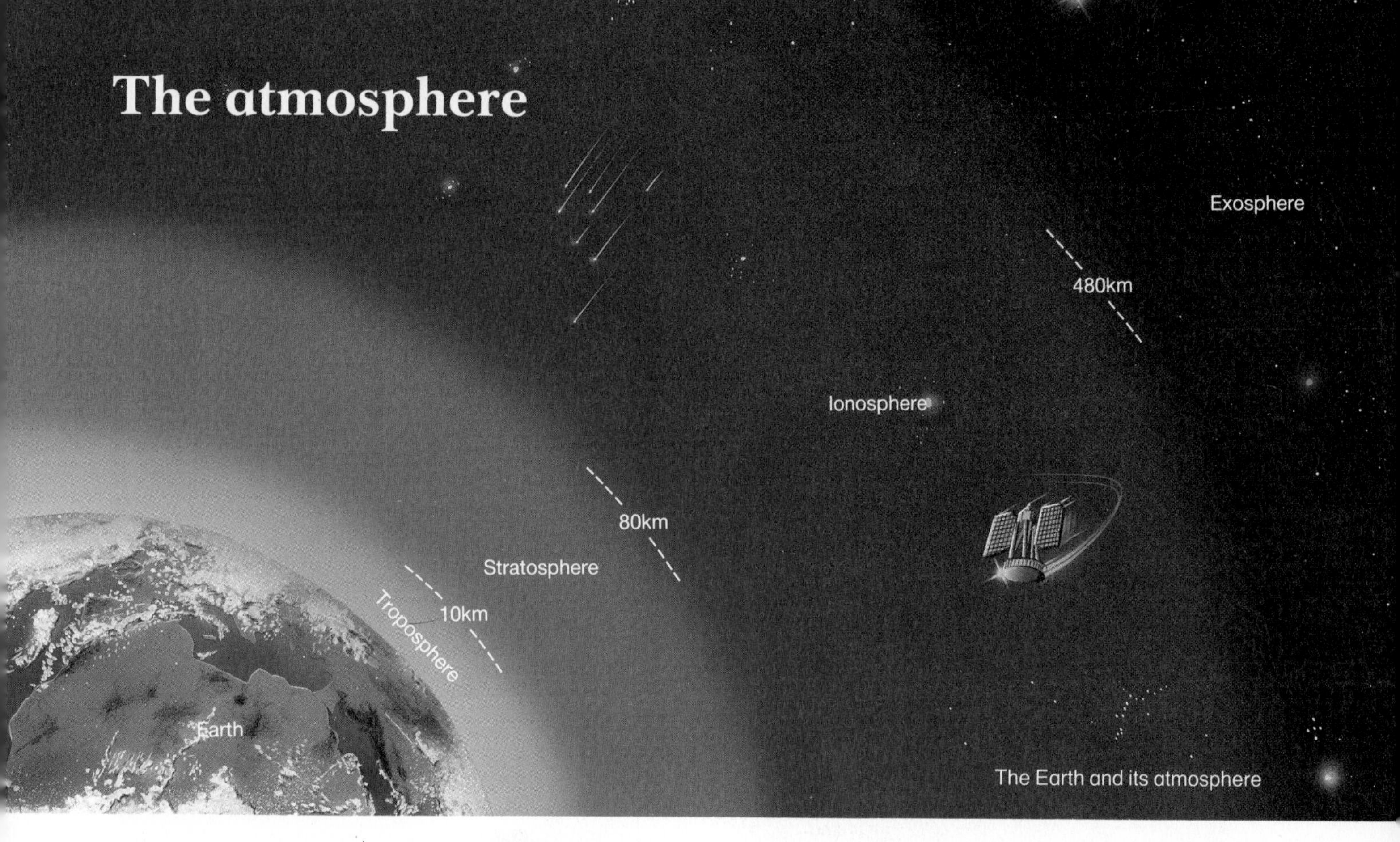

The Earth and its atmosphere

All around the Earth is a big layer of air. This layer of air is called the atmosphere. At ground level there is plenty of air. As we go higher and higher there is less air. Hundreds of kilometres up in the sky there is no air at all. Then we are in space.

No animal, large or small, can live without air. We humans, like all other animals, must breathe air to live. Air has several gases in it. The most important of these gases is oxygen.

The atmosphere is rarely still. Usually the air is moving somewhere over the world. We call these movements of the air wind. Light winds are called breezes. Stronger winds are called storms and gales. The air moves because it has been warmed indirectly by the sun. The sun's rays warm the land. The land warms the air above it, and the warm air rises. The warm air rises because it is pushed up by the cold air around it. Cold air moves in to take its place. This is happening all the time somewhere in the world.

How winds are formed

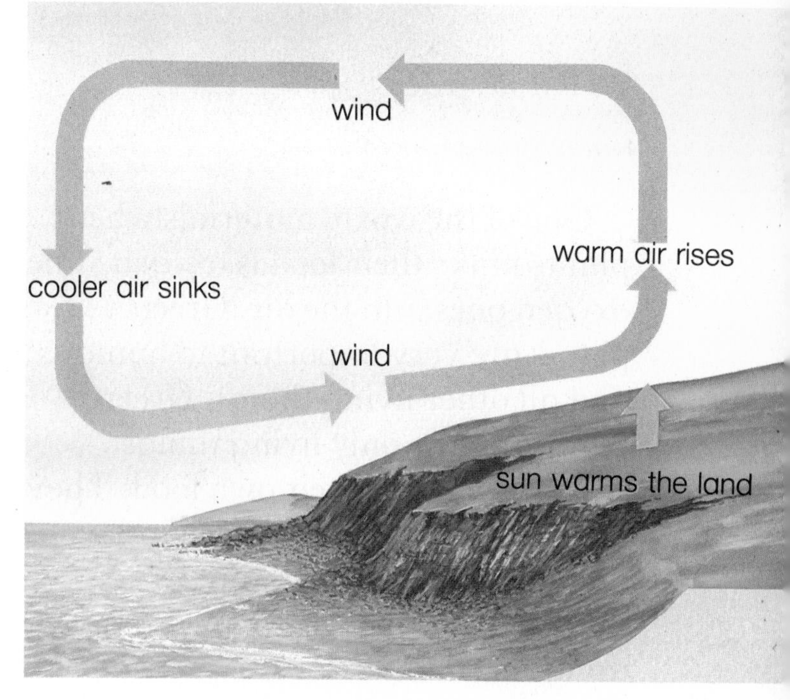

Plants and the air

Like all living things, plants breathe. They take in air through tiny holes or pores in their leaves. But plants also make food in their leaves using part of the air. To make food, leaves need water and mineral salts from the soil. Leaves also need sunshine and a gas called carbon dioxide, from the air. The green substance in leaves is called chlorophyll. This green chlorophyll uses sunshine to turn the water, mineral salts and carbon dioxide into food for the plant.

Trees are vital in towns

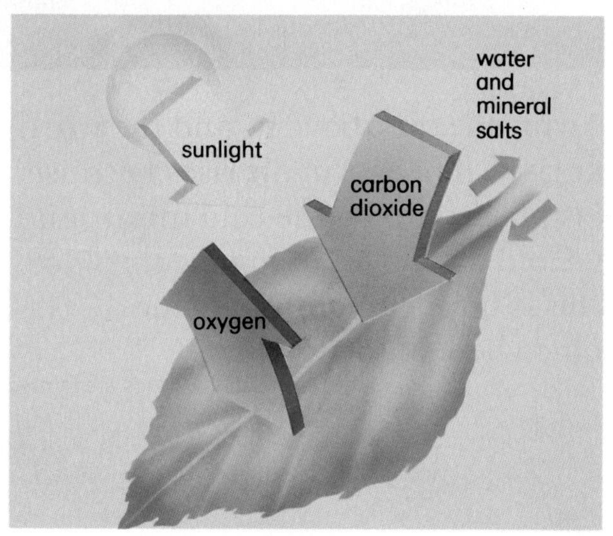

How a plant makes its food

One of the waste materials when plants make their food is oxygen. The oxygen goes into the air. Green plants are very important to humans and all other living things. Green plants are the only living things which can make their own food. They provide food for humans and other animals. In sunlight plants also make the oxygen which all living things need.

Harvesting vegetables

Without green plants, the oxygen in the air would soon be used up. Green plants are the only living things which can make oxygen. That is why it is important, particularly in towns, to have plenty of trees and other plants. It is also one of the reasons why the huge forests of the world are so important.

Weather

A foggy landscape in Sussex, England

All over the world, water is turning into the invisible gas water vapour. The sun shining on the oceans and seas turns a lot of the water into water vapour all the time. The water vapour disappears into the air. High in the atmosphere it is cold and the water vapour cools to form clouds. The millions of tiny drops of water which make a cloud are so small and light that they float in the air. If they are cooled still more, as when the clouds have to rise to pass over a mountain, the tiny drops of water join together. The big drops which are formed are too heavy to float in the air. They fall to the ground as rain or hail.

A rainstorm in London

Clouds may be formed right down near the ground. Then we have mist or fog. You can see the tiny drops of water on your clothes if you go for a walk when it is misty or foggy.

When the weather is very cold, the tiny drops of water in the clouds may turn to ice. Each little piece of ice forms a shape called a crystal. The ice crystals grow bigger and fall as snowflakes. Snowflakes form very beautiful patterns. Rain, snow, hail, mist and fog are just some of the changes we call weather.

Snowfall, and a magnified snowflake

Dew, frost and ice

Often we wake to find that the grass, leaves and stones are covered with small drops of water. This is called dew. The water which made the dew was in the air. When the air is warm we do not notice the water vapour in it. But at night, when the air gets colder, even in summer, it can no longer hold as much water vapour. And so some of the water vapour settles on the cold surfaces of grass, leaves and stones forming small drops of water. This is dew.

If the air is very cold, as it often is in winter, then the tiny drops of dew freeze to form frost. The white powder we call frost is, of course, ice. In cold weather ice may also form on wet roads and puddles. This ice is dangerous to traffic as it can cause accidents.

Dew on wheat plants

The surface of water freezes first. It is a good thing for water plants and animals that ice forms on top of the water. If ice formed at the bottom, the plants and animals would be frozen to death. Fresh water freezes more easily than salty sea water. That is why ponds, lakes and rivers often freeze in winter. The temperature has to be very low for the oceans and seas to freeze.

Frost-covered ivy plants

Climate

Climate is the average weather in a particular place. It is the kind of weather a place is most likely to have at any time of the year. Different parts of the world have different climates. There are several reasons for this. One is the distance of a place from the Equator. The further we go away from the Equator, the less heat there is from the sun. This is because the sun is not so directly overhead.

Snow-capped Mount Egmont, New Zealand

The distance of a place from the sea also affects its climate. Both Edinburgh and Moscow are about the same distance north of the Equator. But the winters in Moscow are much colder than those in Edinburgh. Moscow's summers are also usually hotter than those of Edinburgh. Edinburgh is not so hot or as cold as Moscow because Edinburgh is near to the sea. The sea

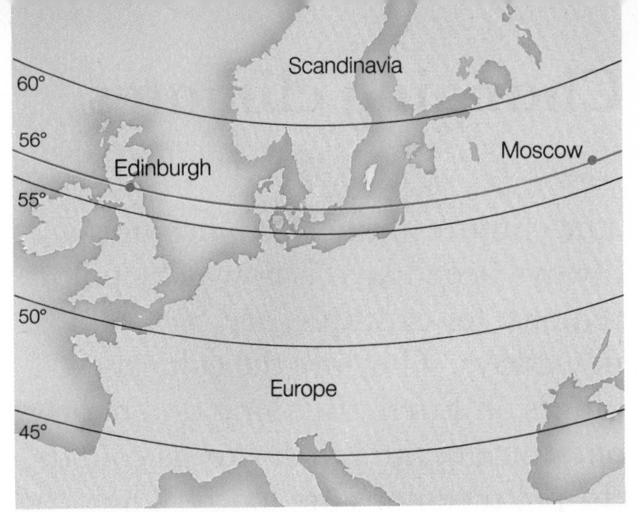

The location of Edinburgh and Moscow on latitude 56°

gets warm slowly but it also loses its heat slowly. As a result, the winds which blow from the sea towards Edinburgh are never very hot or very cold. Warm or cold ocean currents can also affect the temperature of places near the sea.

Moscow in summer

Another reason for the differences in climate is the height of the place above sea-level. The temperature falls about 1°C for every 150 metres you climb up a mountain. That is why high mountains are often capped with snow. Mountain areas also tend to be wetter than lower land. When winds bring moist air towards mountains that air has to rise. The air becomes cooler and the moisture may fall as rain or snow.

Changing climates

The climate of the Earth has not always been like it is now. The polar regions, for example, are very cold nowadays. They are the coldest places on Earth. But long ago nearly one-third of the Earth was as cold as the polar regions are today. The Earth was in what is called the Ice Age. Actually there were at least four of these Ice Ages. In between there were warmer spells.

A scene in Antarctica

During the Ice Ages all of Canada and much of the United States and northern Europe were covered with ice. In the south, the Antarctic ice-sheet was also much larger than it is today. When the Earth became warmer, about 10 000 years ago, much of the ice melted. Where glaciers had been, U-shaped valleys were left. Some of these valleys filled with water, forming lakes. The low hills of clay and pieces of rock which

U-shaped valley on the Isle of Skye, Inner Hebrides

had been pushed along by the glaciers now form valuable farmland. With the melting of the ice, the sea-level rose. Some of the land was flooded.

Even the climate of the polar regions has probably changed. One reason for believing this is that coal has been found in the Antarctic. Coal was made from trees and other plants which grew millions of years ago. The Antarctic must have been warmer for these plants to have grown.

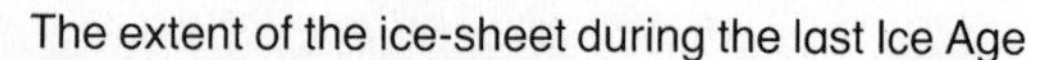

The extent of the ice-sheet during the last Ice Age

The water cycle

There is only a certain amount of water on Earth. This water is being used over and over again. It is being used repeatedly in what is called the water cycle. The energy to power the water cycle comes from the sun.

As we have seen, most of the Earth is covered by water. Nearly three-quarters of the surface of the Earth is covered by oceans and seas. There is also a lot of water on the land in lakes, rivers, ponds, streams and in the soil. This water comes from rain.

All over the world, as we have seen, water is being turned into water vapour by the sun's warmth. High in the sky this water vapour forms clouds. The tiny droplets of water in the clouds may join together and fall as rain, hail or snow. The water in the rain, hail or snow passes into lakes, rivers, ponds and streams and into the soil. Much of the water eventually finds its way back to the oceans and seas. It is carried there by rivers and streams. Or the water may go straight back to the air as water vapour.

The water we drink comes from lakes, rivers, streams or wells. It is usually made clean and pure before we use it. The dirty water from our houses, shops and factories may be put back into rivers or the sea. There it becomes part of the water cycle again.

The filter tanks at a sewage works

An average household's daily use of water

The water cycle

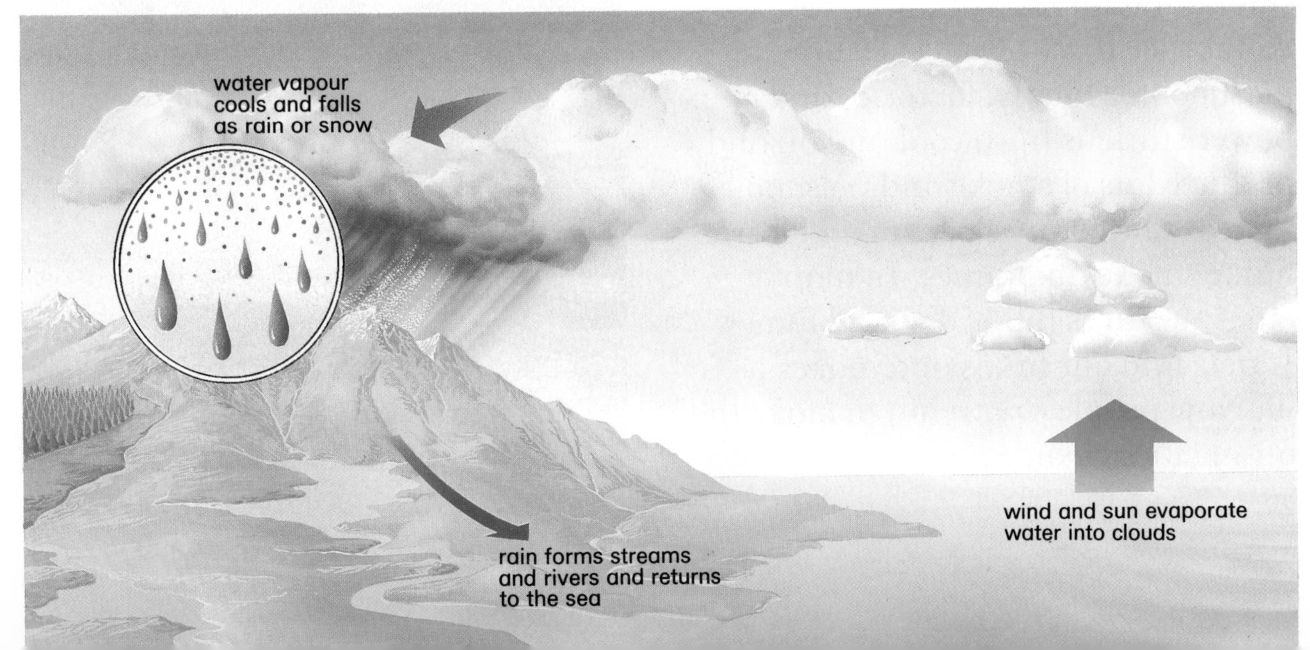

Pollution

Two of the most important substances in our life are air and water. Without air and water we would quickly die. In the country, the air is clean and pure. City air is much dirtier. We say that the air is polluted.

Air pollution from factory and household chimneys

When anything burns it gives off smoke and fumes. Although we cannot see these fumes they are poisonous. Houses, factories and power stations burn coal and oil and produce lots of smoke and poisonous fumes. The engines of cars and lorries make smoke and fumes. In many cities the air is full of dirt and fumes. It gets into our lungs and makes us ill. Now people are trying to make the air in cities cleaner.

Many towns and factories are built by lakes and rivers so that they can use the water. But this water is not always cleaned before it is put back in the rivers and lakes. The water may even contain poisonous substances. Chemicals put on farmland to kill insect pests and weeds and to make crops grow better may be washed into rivers and lakes. All this dirtying or pollution of water makes rivers and lakes smell. It kills water plants and animals. It makes the water unfit for human use.

Even the sea becomes polluted by substances dumped in rivers. The sea is also polluted by oil and other chemicals from ships. Some seaside towns pump their sewage straight into the sea. This is unpleasant for bathers, and spreads germs.

Clearing up oil pollution in the sea

Polluted river water

People all over the world

The children on the left live in Nepal. The girl in the centre lives in Africa and the girl on the right lives on the Isle of Harris, Scotland.

Of all the millions of living things on Earth, the most remarkable are we humans. We developed or evolved over thousands of years from an ape-like animal. The first people probably lived in Africa. In time they spread around the world.

People in different lands evolved in three main ways. White or Caucasian people evolved pale skins to help them to absorb the weak sunlight in the countries where they lived. Negroes have dark skins to protect them from harmful rays in the hot sun. Mongoloid people have stocky bodies and short arms and legs so that they lose less heat from their bodies in cold climates. The early people developed language. They used their large brains and nimble fingers to make tools. They learned how to feed, clothe and house themselves using materials from their surroundings. They discovered how to make fire to keep themselves warm and to cook their food. They invented the wheel. This made it easier for them to move themselves and their goods about.

Other methods of transport allowed people to spread still further across the world. Now there are people almost everywhere. People can live in the hot tropics and the bitterly cold polar regions. They can live in burning deserts, in hot, wet tropical forests and high in the mountains. People have even lived for a short time on the bottom of the sea.

The people of the world can all be friends

Spaceship Earth

In some ways our Earth is like a spaceship circling in space. Like a spaceship, our Earth has room for only a certain number of people if they are to live in reasonable comfort. It has only so much air, water and food. There is only a certain amount of fuel such as coal, oil and natural gas, and of the useful minerals.

But we are using up the air, water, food, fuels and minerals at an ever-increasing rate. One reason is that there are more and more people living on our spaceship Earth. By this time tomorrow there will be 200 000 more people on the Earth. By this time next year there will be about 70 million more of us. All of these people need air to breathe, and food, water and clothes. They need somewhere to live.

At the same time we are polluting the Earth. We are polluting the air and water so that it is no longer fit to use. We may even be changing the climate of the world by cutting down the large forests. We are covering much fertile soil with concrete. By our carelessness we are allowing precious soil to wash or blow away. This means it will be even harder to feed and clothe the world's growing population. From now on we shall have to be much more careful in the way we treat our spaceship Earth.

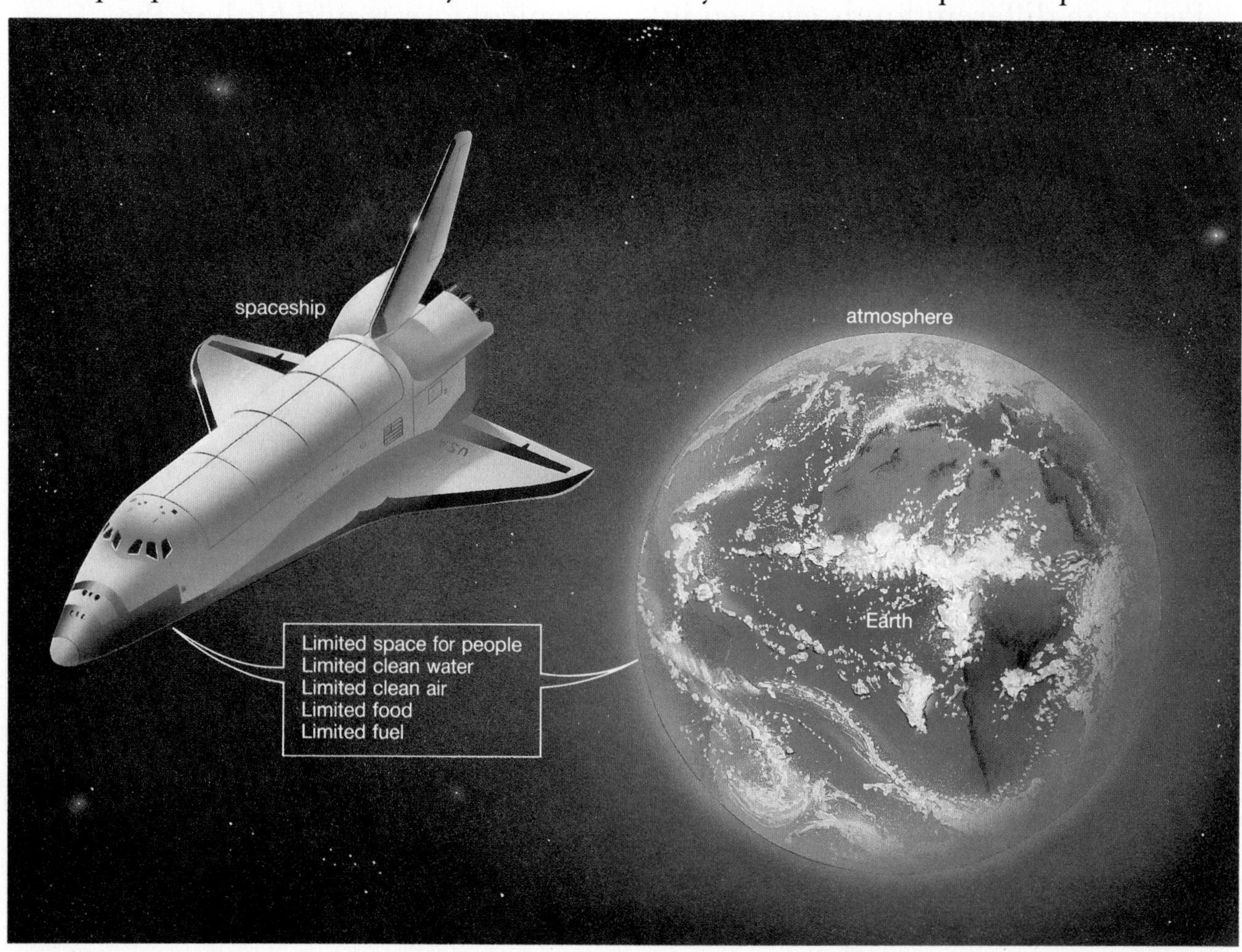

Do you remember?

1 Roughly how much of the surface of the Earth is covered by oceans and seas?

2 Name the four great oceans.

3 How did the salt get into sea water?

4 Which is the saltiest sea in the world?

5 How are waves caused?

6 What is the main cause of tides?

7 What causes most ocean currents?

8 What is the gently sloping platform of land around the coasts of the continents called?

9 What is the abyss?

10 How were the mountains under the oceans and seas formed?

11 What is the name given to the big layer of air around the Earth?

12 What is the most important gas in the air?

13 What makes the air move and causes breezes, storms, gales and other winds?

14 What do plants use to make their food?

15 Why is it important to have plenty of trees and other plants in towns?

16 What happens to water when the sun shines on it?

17 What has to happen to clouds to make them give rain?

18 How are snowflakes formed?

19 How is dew formed?

20 Why is it a good thing that ice forms on the top of water first?

21 What is meant by the word *climate*?

22 What are three reasons why different places have different climates?

23 What happened to the sea-level at the end of the Ice Age?

24 Why do scientists think that the Antarctic was once warmer than it is now?

25 What is the water cycle?

26 Where does the energy to power the water cycle come from?

27 What do we mean when we say that air or water is polluted?

28 What are two ways in which the sea may become polluted?

29 Where did the first people probably live?

30 What are the three main groups of people called?

Things to do

1 A seashore collection Make a collection of things which have been washed up on the seashore. Some of the things you might include in your collection are seashells, seaweeds, dried dead starfish, crab shells and claws, and other parts of dead plants and animals. Wash the things you have collected in tap water. Dry them and make an exhibition of them in your classroom.

You could also make a separate collection of litter washed up on the beach (litter is a kind of pollution). You should *only* do this with the help of a grown-up, though, in case any of the litter is dangerous. It is not unusual, for example, for dangerous chemicals to be washed up on beaches.

2 How rain is formed from sea water This simple experiment will show you how clouds and rain are formed when water in the oceans and seas evaporates. Ask a grown-up to help you.

Half fill a kettle with water. Add about a tablespoon of salt. This is your sea water. Heat the kettle until steam comes from its spout (steam, like clouds, consists of tiny drops of water). Hold a saucepan of cold water near the steam. Collect the 'raindrops' in a saucer placed underneath. Turn off the heat after a few seconds so that the kettle does not boil dry.

If you use an electric kettle *your hands must be dry* before you touch the switch.

3 Seaweed and the weather Many people say that you can use a piece of seaweed to tell what the weather is going to be like.

Find a good-sized piece of seaweed. Take it home and hang it in a garage, porch, shed, greenhouse or conservatory. Is the seaweed wet before rainy weather? Is the seaweed dry before a period of fine, dry weather?

4 Oxygen in water Animals living in water, like those on land, must have the gas oxygen to breathe. Some animals obtain this oxygen from the air at the surface, others obtain oxygen which is dissolved in water. Plants also give off oxygen when they make their food. Plants need sunlight to make food and to produce

5 What does light do to plant seedlings? Put some cotton wool in each of three saucers. Sprinkle a few cress seeds on the cotton wool in the saucers. Water them.

Stand one saucer in a dark cupboard. Put another on a sunny windowsill.

Cut a small hole in one end of a small cardboard box about the size of a shoe box. Cover the cress seeds on the third saucer with the cardboard box.

Look at all three lots of seeds every day. Water them every day. What differences do you see? What does light do to plant seedlings?

6 How heavy is air? Find an old ruler that is no longer of any use, or a piece of thin wood which is about the same size as a ruler.

Place the ruler on the edge of the table so that the middle of the ruler is just on the table. Strike the part of the ruler which juts out with the side of your hand. What happens?

Now put the ruler on the table again in the same position as before. Carefully cover the piece of ruler which is on the table top with a sheet of newspaper or thin card. Again strike the part of the ruler which juts out with the side of your hand. What happens? Why is there a difference? Is it the weight of the paper which has made the difference?

7 Find out about evaporation

Find out about evaporation. Get a saucer and a small meat paste jar. Fill the jar to the top with water. Carefully empty this water into the saucer. Stand the saucer on a windowsill or shelf indoors. Fill the paste jar to the top again with water. Stand it by the side of the saucer of water. What happens to the water after a day or so? Which dries up first? Why? Where has the water gone?

8 What happens when air is heated? On page 33, we saw that the sun warms the land, and the land warms the air above it. You can see what happens to air when it is heated if you do this simple experiment. You will need a plastic bucket, a plastic bottle and a balloon.

Cool the plastic bottle by standing it in cold water or in the refrigerator. Stretch the neck of the balloon and fit the balloon over the neck of the bottle in the way shown in the picture.

Now stand the bottle in a bucket of hot (*not boiling*) water. This will warm the bottle and the air inside it.

What happens? Why do you think this is?

Things to find out

1 Why is there no weather on the moon, even though the sun shines on the moon just as it does on the Earth?

2 Make a list of as many 'weather-words' as you can find – sunny, cloudy, windy, dull, and so on.

3 Do you think that small insects living in a tussock of grass would experience the same climate as cows living in the same grassy field? Say why.

4 How many reasons can you think of why people like to know what the weather is going to be like in advance? How might the weather forecast help some people, such as fishermen, farmers, pilots and sailors, with their work?

5 Find out what a barometer measures. How does it work? What can we learn about the weather by studying the readings on a barometer?

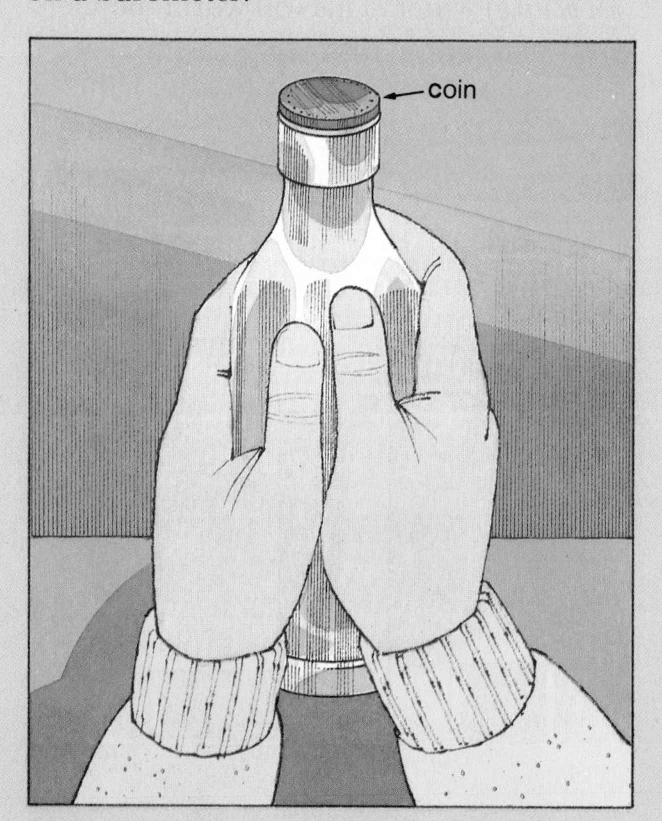

6 Here is a simple but fascinating experiment. Put an empty lemonade or wine bottle in the refrigerator for a few minutes until it is cold. Then sprinkle several drops of cold water around the top of the bottle. Now sprinkle a few drops of cold water on a coin.

Put the coin of top of the bottle. Put your hands around the bottle, covering as much of it as possible. After half a minute or so the coin will move. It may go on moving even when you take your hands away from the bottle.

Try to find out why the coin moves. You may find that reading page 33 again helps you.

7 The Beaufort Scale is often used to estimate the speed of the wind. Find out how the Beaufort Scale works and who invented it.

8 What is the difference between mist and fog? Sometimes 'smog' may occur. What is it?

9 Study the weather forecast in a newspaper, on the radio or on television, and then see what the weather is really like the next day. How often is the weather forecast right? How often is it partly right, or totally wrong?

10 Our Earth can be divided up into three broad regions according to the climate they have: the tropics, the temperate regions and the polar regions. Find out what kind of climate each of these three regions has. What are the trees and other plants like which grow in these regions?

11 Find out the names of some countries – large or small – where the weather remains almost the same throughout the year.

12 It is sometimes said that a place or

country has extremes of climate. Find out what is meant by this. What places or countries might be said to have extremes of climate?

13 Why do so few people live in the deserts and the polar regions? What could be done to encourage more people to live in these places?

14 Why is the eastern side of Britain drier than the western side?

15 On page 38 we read that coal has been found in Antarctica. This suggests that Antarctica was once much warmer than it is now. When do you think that the Antarctic was warmer and why do you think it became colder?

16 Why is Antarctica colder than the Arctic regions?

17 Some desert countries are near to the sea. How could these countries obtain drinking water, and water to make plants grow, from sea water? Why is it not done?

18 What is done to help to get rid of patches of oil floating on the sea and washed up on beaches?

What can be done to help seabirds whose feathers have got oil on them?

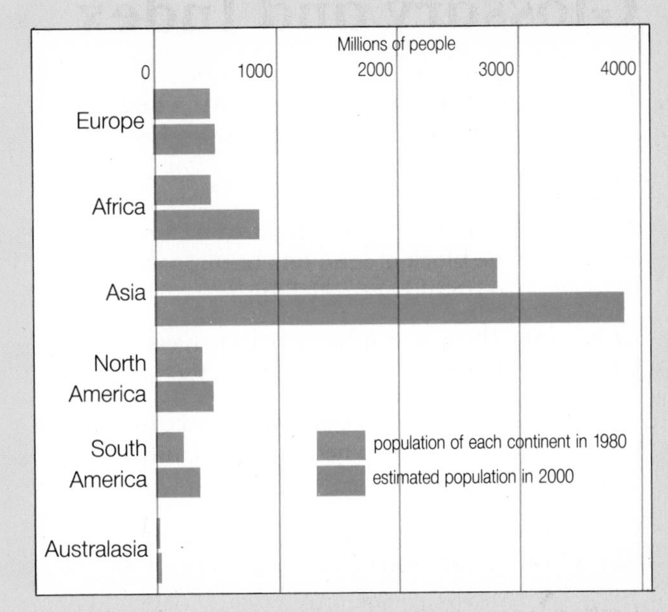

19 What kinds of pollution do you come across regularly? What could be done to stop these kinds of pollution?

20 Find out how pollution of the sea affects coral and coral islands. What other things threaten the living coral?

21 One kind of pollution is called 'acid rain'. What is acid rain, how is it caused and what harmful effects does it have?

22 How can we make the world's supplies of fuels, minerals and metals last longer?

23 Which countries of the world are very overcrowded?

24 It is believed that more than half of the people in the world do not have enough food to eat. In many countries there are serious famines from time to time and people die of starvation. Find out the names of some of the countries where there are serious food shortages and what causes them. What, if anything, is being done to help the people in these countries?

25 Describe what life would be like if there were twice as many people living in the area where you live.

Glossary and Index

Here are the meanings of some words which you might have met for the first time in this book.

Abyss: the deepest part of the oceans.

Block mountains: flat-topped mountains formed where a large block of land has been pushed up between two roughly parallel faults.

Chlorophyll: the green substance in leaves. Chlorophyll helps plants make their food.

Continental shelf: a flat or gently sloping area of land which forms a border to nearly every continent. The continental shelf is covered by sea water.

Continental slope: the steep, cliff-like slope from the edge of the continental shelf to the ocean's depths.

Fault: a large crack or break in a series of rocks.

Fold: the bending of rocks caused by movements of the Earth's surface.

Hemisphere: half a sphere. The Earth is divided by the Equator into a northern hemisphere and a southern hemisphere.

Humus: the decayed plant and animal remains found in the soil.

Ice Ages: the periods of time, thousands of years ago, when the Earth was much colder than today.

Igneous rocks: rocks formed out of lava.

Lava: the hot, liquid rock from a volcano.

Meander: a bend in a river.

Metamorphic rocks: igneous or sedimentary rocks which have been changed by heat or pressure in the Earth.

Polar regions: the very cold areas around the North and South Poles.

Sediments: particles of mud, sand, gravel and other pieces of rock which sink to the bottom of water.

Sedimentary rocks: rocks such as chalk, clay and limestone formed by the hardening of sediments.

Stratum: one layer of sedimentary rock. If there is more than one layer, the word strata is used.

Tides: the rise and fall of the level of the seas and oceans caused mainly by the pull of the moon's gravity.

Volcano: a weak part of the Earth's crust out through which molten rock, or lava, flows.